Automation in a polytechnic library: fifteen years' development at Hatfield

D E Bagley and E Oyston

Case studies in library automation

LA

The Library Association · London

First published 1982

British Library Cataloguing in Publication Data

Bagley, D. E.
   Automation in a polytechnic library.—
   (Case studies in library automation)
   1. Hatfield Polytechnic. *Library—*
   Automation
   I. Title   II. Oyston, E.   III. Series
   025'.0028'54   Z769.9

   ISBN 0 85365 964 8

Designed by Geoff Green

Typeset by Input Typesetting Ltd, London SW19 8DR

1 2 3 4 5 6 86 85 84 83 82

Contents

List of illustrations

Series note

The Library Association series of *Case studies in library automation* provides for librarians in service and for students of librarianship a readily accessible source of information on the practice of library automation.

The subjects of the case studies have been chosen to reflect the diversity of applications that has evolved in the last ten years, and the wide range of libraries that have successfully carried through automation projects. The series will eventually cover academic libraries (both university and polytechnic libraries); public libraries; special libraries; and library co-operatives. The computer technology will range from locally developed special purpose systems to integrated 'package' solutions provided by library automation suppliers.

The authors of the books in this series are, with one exception, senior librarians who have taken major responsibility for the projects they describe. The studies are, therefore, written from the librarian's point of view and care has been taken to ensure that computer 'jargon' is used as little as possible (and explained, when it is unavoidable), and that the accounts should be comprehensible to any librarian or student with a minimal knowledge of what computers may be expected to accomplish.

Authors have been encouraged to develop the reasons for the decisions to commit to automation and the selection of their particular course of action; to describe the stages of implementation; and to appraise the lessons learned from their experience and the eventual effectiveness of the new systems.

Each case study is complete in itself and the cost has been held to a minimum in the hope of keeping it within reach of

2   students and newly qualified librarians. The texts are intended
to be suitable either for private study or as supporting material
for courses in library automation.

The first study in the series, *Studies in the application of free text
package systems in libraries and related information centres*, differs
from the subsequent volumes in that it not only contains a
number of cases drawn from special libraries, but also provides
a general review of library automation as it now appears and
sets a framework for all of the later accounts.

The present study, *Automation in a polytechnic library: fifteen
years' development at Hatfield*, follows the development of library
automation in a new polytechnic from its earliest days to its
eventual incorporation in a successful co-operative system for
cataloguing and circulation control.

The Hatfield Polytechnic Library is unusual amongst academic libraries in that it functions also as the headquarters of
a county-wide network of libraries in smaller colleges and provides an information service to industry. The need to provide
central supporting services gave impetus to the introduction of
automation in the Library, and the rapid increase in both load
and resources arising from designation of the original Hatfield
College of Technology as a Polytechnic added urgency to the
early initiatives.

This case study falls into two main sections: Part 1 records
the early planning, the use of British Library systems as they
developed, and the use of 'in-house' computing facilities as a
low priority user. The second part describes in some detail the
reasons for choice of SWALCAP as the eventual preferred system, the tasks of conversion, including the transfer and reclassification of 80,000 computer records, and the review at that
time of its 20 years of progressive automation, and redesign for
the systems available in 1981.

Features of particular interest are Hatfield Polytechnic Library's longterm and determined adherence to the principles of
library co-operation; the willingness to cost and evaluate each
successive opportunity as library automation developed; and
the patience and persistence of the Library staff in seeing
through each stage of the long evolution. It provides a valuable
account of computer development in an institution where *everything* was new, by design; where automation was treated as
'normal' as early as 1966; and where co-operation between the
Polytechnic Library, smaller colleges, local industry, and the
County Public Library was a matter of course.

The Hatfield Polytechnic Library's view – as a member – of   3
the services of LASER and SWALCAP complements the
accounts of these systems which themselves are the subjects of
Case Studies in this series.

J.H.A.

Part one: Fifteen years development

1
Hatfield and HERTIS

In order to see the progress of ideas at Hatfield it is first necessary to understand their context because the Hatfield Polytechnic library since its inception has had a role wider than its own academic establishment.

The provision of college libraries in Hertfordshire was based on planned and co-ordinated development, mainly in the period 1955–65. This coincided with the post-war development of Hertfordshire, changing the county from an essentially rural community to one which took a good deal of London's overspill into five New Towns. The Abercrombie plan for Greater London advocated population moves to new areas in the Home Counties, and apart from Crawley and Harlow, Hertfordshire took the brunt of this development. As a consequence, however, the County Council had both incentive and finance available to embark on a new dimension of planning at all levels. The Education Committee, in particular, was influenced by the White Paper of 1956 on Technical Education[1] which encouraged provision for the education of craftsmen, technicians and scientists.

Hertfordshire drew up plans for a greatly increased population which included the development of some 15 establishments of further education (Fig 1). Although there had been small colleges in the two original centres of population, Watford and St Albans, development which was to have a bearing on college library growth commenced with the building of a totally new Technical College at Hatfield in 1952. Hatfield is conveniently placed in the geographical centre of the county and the master plan for technical education envisaged general provision in Colleges of Further Education in all the main towns, with more advanced work concentrated at Letchworth and Watford, to the

North and South ends of the County, and the highest level specialized work to be carried out at Hatfield. Indeed, only ten years after its opening the Hatfield Technical College was designated one of 25 Regional Colleges and thus became a national centre of advanced education. The nucleus of a library was quickly established by its first librarian, Mr. O M Argles, but major development took place after 1956 with the appointment of G H Wright as County Technical Librarian charged with the responsibility of *developing a network of libraries in the new colleges as they were built.*

Such large scale development of technical education provision in one county was unusual in Britain but because everything was being created from scratch, the pattern of college library development could be carefully planned and integrated. The County Education Committee accepted the idea in the 1956 White Paper that all colleges should have a library, and that they might make a contribution to the provision of information to local industry. Thus the acronym HERTIS was coined to designate the Hertfordshire Technical Information Service, based on an integrated network of libraries in further education establishments (Fig 2). Close working relations were maintained with the County Library system, which, under the Education Committee, provided services to all but four independent borough areas in the County, but the funding for HERTIS was separate and disbursed under Gordon Wright's direction. A redistribution of committees followed the Public Libraries Act of 1964 when the County Library became responsible for the independent areas and reported to a separate Libraries Committee whilst HERTIS remained the responsibility of the Further Education Sub-Committee of the main Education Committee. Under the later reorganization of Local Government (1974), the Libraries Committee was subsumed into a Cultural & Recreational Facilities Committee (Fig 3).

For the first ten years of HERTIS development work concentrated on the creation of libraries in the new colleges, with high priority being given to their educational role within the college and to the value of 'tutor librarians'. The information service to industry was developed with a Chief Information Officer being appointed at Hatfield to act as a salesman promoting the service. When the Ministry of Technology developed its Industrial Liaison Officer scheme soon after, Hatfield and HERTIS were regarded as a model for the operation. Because of the network of libraries, attention was centred on the problems of document storage, retrieval and transmission, and Hatfield became a centre of expertise concerning microfilms and filming, and document reproduction methods and machines. At an early stage research was being carried out into what could have become computerized retrieval linked to microform stores using a Filmorex machine, a relatively primitive device capable of automatically sorting chips of microform selected by keywords. This work led later to an OSTI contract to survey use of microforms in the UK,[2] and hence to the creation at Hatfield of the National Reprographic Centre for documentation (NRCd) in 1967, one of the earliest and most successful of what have become British Library's special information centres.

With the growth of the library network came the opportunity to apply various forms of automation to the housekeeping side. In the interests of library co-operation, book purchasing was linked to the County Library's system. A single county-wide accession number sequence was used, with blocks of numbers allocated by County Library HQ. College libraries then ordered their own material, passing order slips through the nearest Regional Headquarters Library to ensure adequate local knowledge of overlapping interests. (At that time the County Library operated in five Regions, but later concentrated into three larger Divisions.) Because of their more specialized interests, the three largest colleges, Hatfield, Watford and Letchworth passed their records directly to the County Library main headquarters. The Colleges of Further Education collected stock related to their specific course needs, whilst catering also for general education and for evening class interests. Their ordering was influenced by entries appearing in the weekly British National Bibliography (BNB) and it proved convenient to obtain BNB cards for them via the Regional or Divisional County Library offices. BNB practice regarding classification was also considered generally acceptable, except in the three specialized colleges. Consequently the Universal Decimal Classification (UDC) was adopted for these colleges and catalogue cards were provided centrally by Hatfield, where a printing machine using embossed metal plates was installed (Fig 15).

This equipment represents the first major introduction of automation since it was used centrally to produce multiple copies of bibliographic records on cards, run off on to rolls of card stock and fed through a guillotine to provide 5 x 3 cards. Some of these would be for the library catalogues but in addition a good deal of scanning of technical journals took place to provide a current awareness service to industry, based on batches of cards in classified sequences containing brief descriptive annotations, an early form of SDI provided by a central independent agency.[3]

2

Early priorities for automation

The author was appointed as Deputy County Technical Librarian in January 1966, with a brief to consider further automation as an explicit part of his job description. At that time, use of computers for library applications was largely confined to the production of indexes, KWIC being a popular form, although a few universities were researching into large file handling problems, notably at Newcastle.[4]

Within HERTIS it seemed clear that early consideration should be given to the problems at Hatfield, headquarters library for the network and already the biggest of the libraries in terms of student numbers and intensity of use. The priorities were to consider eventual replacement of the card production equipment, and possible mechanization of the necessarily complex issue system. In April 1966 the Hertis Advisory Committee was given an outline report of the potential application of computers to library systems and information retrieval, and noted a recommendation that a paper-tape typewriter be considered for purchase in 1967.

At this time the Hatfield College owned an Elliott 803 computer, an early but effective machine with magnetic film memory, and a student project was organized to consider the feasibility of compiling and maintaining batch production of a subject index. Progress was made, but such are the problems of student projects that although the programs were written to *compile* entries, the student left before completing the amendment and deletion programs! However, useful lessons were learnt from this and subsequent attempts to keep a computer-produced index up to date, and in 1969 a package developed by City University in conjunction with a computer bureau was adopted. This produced UDC subject indexes until 1976 but was never considered satisfactory because of high costs and slow production.

An indication of the national state of library computerization at this time can be gained from the Parry Report on University Libraries (1967) which included a section on the relevance of computer operations for libraries, recommending that library needs be borne in mind when computer systems were upgraded or replaced. Unfortunately the UGC Computer Board preferred to think only in terms of research needs and would not countenance guarantees of computer time for administration purposes. The SCONUL report of 1973[5] by Norman Higham underlined and quantified these needs, and the British Library ADP study of 1972[6] listed a number of criteria for computer-based operations, eg:

a The new system should provide a better service at lesser or no greater cost.

b The new system should give added benefits at lesser cost than would be provided by present systems.

c The new system could give added benefits and services and recover additional costs from consumers of the benefits and services.

d The new system could contribute towards larger (national) systems in ways recognized as being desirable but at present not achievable economically (or technically).

These are fundamental words of wisdom which should always be borne in mind!

In 1968 the then Hatfield College of Technology underwent another metamorphosis through its designation as a Polytechnic, following the Polytechnic White Paper of 1966. (See chronology, figure 15). This change meant a major increase in size and gave greater urgency to rethinking the library systems. Within two years the number of students doubled to 3,000, with an anticipated target of 6,000, and the library bookfund quadrupled – but with little increase in staff! An extra site was acquired, eight miles away, and plans were agreed for a new multi-storey library block as the top priority for the physical expansion of buildings on the main site. Despite this sudden growth Hatfield was lucky in one respect by comparison with the majority of the other Polytechnics. It was one of only four out of the 30 institutions which was not required to be formed by amalgamation between several distinct and separately

10   developed constituent colleges. Hatfield was also by far the youngest, having been in existence only since 1952, and had been poised for extension either into a College of Advanced Technology or into a new University of Hertfordshire, both of which had been contemplated but had not come to fruition. Thus considerable expansion had already been considered and a new Development Plan for the Hatfield Polytechnic was very quickly drafted. Part Three of the plan was concerned with the role of a library in a university-scale institution, and it was accepted that the Library would be a major teaching instrument with a central role in the educational programme. Other paragraphs included the concept that it would be a dynamic library service using new techniques to the full, with many routines performed by mechanization and with close links between library and computer centre.

Hatfield was fortunate in having a well developed Computer Science teaching department which lost no time in acquiring a new large DEC-10 system specifically chosen for its ability to provide a county-wide online computer facility for use by colleges and schools. Unfortunately the DES concurred with the UGL Computer Board and specifically stated that large scale use by the Library for administrative purposes should not be encouraged. However, a Computer Centre was created with staff who accepted that in practice, batch work could be undertaken in off-peak time.

3
Issue or catalogue – the need for a database

Although it had been accepted earlier that a computer-driven issue system was economically desirable it had become clear that a catalogue database would have to be constructed first. During an earlier exercise in work simplification, it had been found that the meagre accession records which had been kept, using random batches of accession numbers issued by County Library HQ, were not worth retention. Consequently the only shelf list available was the classified sequence of the catalogue. The issue system at Hatfield was based on three-part slips on which all author/title details were handwritten by the borrowers. One set was filed in accession number order, but there was no way of identifying a book if this number alone was quoted.

At Easter 1970 the Polytechnic Library moved into a new four-storey home, with book-stock spread over three floors representing major sectional interests. A second site had also to be considered where a library would have to be provided for the Business Studies degree and the development of undergraduate studies in Social Sciences. Catalogue production by computer had thus become more urgent because multiple sets of the catalogue were very desirable.

As part of the Academic Plan it was intended that the Polytechnic would develop studies in the Humanities and Arts in fulfillment of polytechnic philosophy of embracing the two cultures, or 'many arts, many skills' as the slogan became. This posed a further and very real problem since the UDC classification used to date could prove unworkable in these new areas.

Bearing in mind that BNB MARC tapes were at this stage still only a long-term prospect, negotiations were conducted with

English Electric Computers Ltd to prepare a library catalogue, on the basis that the Polytechnic's own computer centre would undertake to continue the exercise once a database had been established. A general specification was drawn up for a KDF9 computer to process a combination of fixed and variable length field data, records to be provided initially by Hatfield from a Vonamatic paper-tape machine. Progress as far as agreeing on draft cataloguing work sheets was achieved, but outside events such as the merger of English Electric Computers with ICT Ltd intervened and finally the costs escalated beyond the limited budget available.

The HERTIS Advisory Committee had been told early in 1969 that discussions were about to be finalized on a contract, pending the ability of the Computer Centre to take over the maintenance at a later date, and that 'initial data preparation might also have to be undertaken by an outside bureau'. A further complication, however, was that the Computer Centre although willing in principle was unable to obtain approval for the long term commitment since this would clearly be against previous policy. The next step was to establish the cost of having the catalogue prepared completely externally and Oriel Computer Services were approached. By this time Oriel had successfully demonstrated catalogue production under contract for Belgian university libraries, and could provide exciting KWIC variations of file presentations. Again, however, the quotations proved outside the cash limits available at the time but gave a firmer idea for future budgetting.

The Library occupied its new building in 1970 and the following summer decided to carry out a stock check to establish the reliability of the card catalogue as its basic database. No full check had been carried out in the previous 15 years, including the recent period of very vigorous expansion. Total reclassification was to be considered, but an overhaul of the catalogue seemed a desirable first step. Because of a variety of changes it seemed likely that the author sequence represented the most accurate record, so for shelf checking purposes cards for some 60,000 books were resorted into classified order and the Library was closed for two weeks. In the event, some 20 staff participated, forming themselves into teams, *and the three major subject floors were completely checked within eight days.* The catalogue was skimmed to about 50,000 records considered 'live', and it was then possible to consider whether either full author catalogues or at least microfilmed sequences updated at six monthly inter-

vals could be maintained for each floor and for each site. One 13
master microfilm set was obtained but the overall quality of
frames proved disappointing due to the very variable quality of
card printing. Author sequences for each floor were prepared
by running off new cards from the original metal plates but the
task of maintaining more than one sequence proved too much
and by 1973 a Section Heads meeting agreed that this idea too
must be abandoned.

The Hertfordshire County Library was very interested in the
potential of computer operation to solve some of its problems
and a small Study Group, (including the author on behalf of
HERTIS), had been created in 1967. Its first report in January
1968 concluded that for the County Library, catalogue preparation only was feasible and desirable at that time. Their catalogue
needs were for a minimal entry; fixed fields of 20 characters
were considered to suffice for author and 50 for titles. International Standard Book Number (ISBN) and Dewey Classification
would be incorporated and the entry could be prepared at the
order stage. 50 copies would have been required, with annual
print-outs and fortnightly accession lists cumulating quarterly.
The issue system developed in prototype by Elliott Automation
Ltd (and subsequently marketed by ALS Ltd) had been inves-
tigated, but considered unsuitable since it would have been
slower to operate than photocharging and could not be
borrower-operated. A book order system was considered feasible, but less flexible than the manual methods then in use.
Although cataloguing was considered feasible, the County
Computer Unit was unable to offer time and estimated that
several thousands of pounds of annual expenditure would be
needed. The final conclusion was that more progress might be
made by HERTIS at Hatfield when the Polytechnic had acquired
its new computer, and experience thus gained could be used
later by the County Library.

This report has been quoted in some detail because it em-
phasizes that *both* library systems saw the need to collaborate
on computer developments, but a difference in philosophy
clearly emerges. The colleges felt a need for fuller catalogues in
their libraries because of the different nature of use. Students
required specific books and editions of titles and would use the
catalogue as a selection aid. Academic library issue systems also
need to cater for differing categories of user, allowing varying
numbers of books for different periods. It is also necessary to
know quickly who holds a particular title. The County Library

14   were happy to use BNB as a guide to books available, and thence proceed to the shelf. A stock catalogue could be a minimal record for location purposes only. A public library issue system had to beat the speed of photocharging as a means solely of speeding readers past the issue desk! The later advent of MARC tapes and the success of BLCMP as a catalogue production service led to the re-creation of the Working Group and further ideas in 1976 about computerized cataloguing. However, chronologically one must return to the Polytechnic saga.

By 1971, the Polytechnic had established a pattern of appointing one graduate trainee per year, offering one year of in-service experience followed by a postgraduate course, usually at Sheffield University. By this means a useful variety of people with specific subject backgrounds had been obtained. It was decided that priority should be given to computer problems and an appropriate trainee selected accordingly. As part of his in-service time he took an internal course in systems analysis, and specialized at Sheffield in appropriate areas. On return he was also loaned to INSPEC for a short time to widen his experience but as a result decided that his talents could be better used immediately in the outside world and so severed his contract with Hatfield. The library discovered in this way the difficulty of recruiting and keeping specialist staff in an exciting new field with more competitive salaries outside local government. Subsequently further appointments were made to a new post of library programmer, for staff who would work for the library but in the Computer Centre. Again experience would show that these people were tempted away before long, although valuable progress was made with their full-time commitment.

4
National developments

For the period 1972–74 the only visible signs of progress on the automation front for the library were production of the Subject Index by an outside bureau[7] and infrequent updatings of lists of serials produced in off-peak time by the Polytechnic DEC-10. A major obstacle to greater use of the DEC-10 was the fact that it was a disc-based machine, and no full-size magnetic tape facility was available. Thus, although possible use of MARC tapes was considered, they were technically difficult to handle. Bureau production of the Subject Index was proving costly and never very efficient, and finally the Library made capital available for the purchase of a tape-drive so that the Computer Centre could take over the subject index data. It was intended at this stage that use would be made of PRECIS strings when these eventually became available from the British Library.

A good deal of automation development work was underway in the library world at this time. OCLC was developing in the United States and in the UK BLCMP was beginning to look like a commercial proposition. Work had started at SWALCAP, but this was still considered a research project. In 1969 Gordon Wright had accepted the challenge of moving to Canada to take control of the Ontario Colleges Bibliocentre, an exciting venture providing a central computerized book order and cataloguing facility for a network of college libraries in the State of Ontario. This venture had links with the University of Toronto whose UTLASS programs provided a Canadian alternative to OCLC. The author had useful insight into UK national progress through membership of the British Library Research & Development Department Advisory Committee, and was also invited to participate in LASSOS when this 'think-tank' committee was created early in 1975. LASSOS was intended to provide a small forum representative of various library interests to explore the options for development in the UK of library automation systems, and to explore the possible role for British Library services. The committee's role was seen to be to identify the functions to be served by library automation and the different means of providing for these; and to examine the advantages and disadvantages of various alternatives. LASSOS was kept informed of the likely development of the MERLIN system, then considered a major prospect for a future online central service from BNB. It is relevant to stress the author's involvement in LASSOS and awareness of British Library thinking because this had direct influence on the next stage of development at Hatfield, where the needs for reclassification to Dewey and automated catalogue production had become critical.

The Polytechnic was pursuing its aims of broadening its scope into the social sciences and arts. In 1969 a degree scheme for Applied Social Studies commenced, followed by a part-time degree in Sociology in 1972 and then a wide ranging BA in Studies in the Humanities was launched in 1973 with Philosophy and English Literature as major options. Development of bookstock to meet these needs made a decision on a classification change from UDC essential. Fortunately by 1974 it appeared that the London and South Eastern Library Region (LASER) would soon have completed retrospective conversion of BNB records, with the promise that the BL Bibliographic Services Division could then offer complete retrospective files of MARC standard records, at least for titles included in BNB since 1950.[8] It was expected that these files could be made available in microfilm output form and Hatfield planned to use these for a major reclassification exercise. Studies of the options available indicated that a change to 18th Dewey would be sensible, not only in view of its wide applicability, but also because for the forseeable future this would be the only choice available with centrally produced records.

Since it seemed reasonable to attempt to combine a reclassification exercise with the creation of data for future machine readable records, overlap studies would be necessary to determine the quantity of stock likely to be found in BNB, and hence potentially available in MARC records. Data would also have to be obtained for the duplication of titles between Hatfield and the HERTIS Colleges, and a rough approximation also with the County Library stock. Tentative costings were estimated for using BLCMP services to prepare a retrospective catalogue and for continuing with current cataloguing. Fuller details are given below but it is worth mentioning that the Polytechnic Library also participated at this time with early trials of online information retrieval through a British Library project.[9] Five different kinds of library took part to test the feasibility of accessing a remote computer database, Lockheed in California, to see how easy it would be for untrained users to trace information. Despite the early tribulations (eg telecommunications problems and the very limited experimental database available), these trials showed great promise and further whetted the appetite for eventual online catalogue production and access.

5

BLAISE and LOCAS

1975/6 proved a watershed for library automation in the UK. The BL Bibliographic Services Division provided an experimental catalogue production service with computer output microform for a public library, the BRIMARC project as it become known, with Brighton Public Library as the recipient.[10] LASER carried out a feasibility study on library co-operation for automation, the COLA project completed in mid-1975.[11] The British Library itself announced the creation of BLAISE in May 1976[12] although a brochure announcing the possibility of a full local cataloguing service had first been issued in 1975.

BLAISE could be said to have been born out of the online information retrieval experiments, since it basically adopted the National Library of Medicine Elhill software developed for the MEDLARS database. Information retrieval, especially from MEDLARS, would continue to be offered, but now the MARC tape database would also be available for retrospective searching and, by extension, for agency production for library catalogues. LASER's COLA project stressed the potential value of on-line cataloguing and interlending systems, and later discussions within LASSOS concerned the potential development of regional library computer service agencies. LASER went on to develop and install a minicomputer based system, initially to facilitate interlending, but also with the potential to provide catalogue records (finally fully achieved in 1982).

Given the possible developments at the beginning of 1975, Hatfield decided to plan reclassification and retrospective catalogue conversion linked to British Library services as these became available. LASER might later be another possible local source, because a local London-based service rather than

BLCMP (100 miles away in Birmingham) seemed the rational solution at the time to keep telecommunication costs to a minimum.

The basis for a final decision centred around the likely future development of national plans for library automation. At the beginning of 1975, BLCMP appeared to be a proven, commercial agency able to provide a library with its own catalogue. The BL service however was attractive on paper, and tentative costings for an annual intake of 25,000 titles indicated production costs of an annual catalogue with monthly cumulating supplements to be about £6000.

BLCMP had the advantage of drawing immediately upon MARC tapes plus members' records, and *potentially* would supply records for 85% of likely intake. The BL on the other hand might supply immediately a slightly lower proportion, but it was the central agency with national standing that above all owned the copyright in MARC records.

HERTIS again shared in discussions within the revived County Library Working Party, but it appeared likely that the County's needs could also be met by recovering records from LASER at very low cost since the County was a long standing LASER member. Hertfordshire records would have been notified to LASER over the years, and their requirement would be for a minimal catalogue record only. HERTIS and the County Library agreed to continue on separate lines although ensuring through MARC standards that the separate systems would always be potentially compatible. With this in mind the next important task was to complete the overlap study in order to provide a firmer basis for the costs of obtaining catalogue records and for the local preparation needed for Extra-MARC Material (EMMA data).*

Advice was sought from the Library Management Research Unit, then at Cambridge, on the size of sample required, and finally three years of BNB under the letter J were used. Copies of the relevant pages from annual volumes were used to check against catalogues in each of the HERTIS college libraries. Extrapolation from the sample suggested that Hatfield and the more specialized colleges would have a 50% overlap with the other colleges, although this percentage would fall to nearer 30% for Hatfield alone. It was anticipated that 75% of titles at Hatfield should be in the MARC files, since the Library had

*[See the Case Study in this series: *Computing in LASER: regional library co-operation.* LA Publishing Ltd, 1982. *Ed.*]

been in existence only since the mid-50s and the great majority of its stock would be English language material. Proportions were expected to be higher for the other colleges because of their later development and more limited scope. The County Library also prepared costings for using the British Library cataloguing service and it is interesting to compare these and the different requirements.

HERTIS assumed an annual intake for all colleges of 25,000 titles per year, with 3.5 entries per title for the author/title catalogue. 70% of new material would be in MARC, requiring only the 5p per record charge then being quoted, plus the initial access fee of £250. Local input would cost 60p/1000 characters and averaged 400 characters per record. Computer sorting, file format and COM output figures were based on the published British Library tariff. HERTIS annual costs were estimated at £6,000 per year, for 25 copies of microfilm catalogues, whilst the County Library estimated £4,500 for 40 copies. These figures, equivalent to one salary for professional staff, were sufficiently attractive to encourage HERTIS to proceed further. The County Library was concerned initially that a prior major reorganization of central bibliographic services would be required, and were thus not able to proceed without further discussion of the wider implications. Both agreed at this stage, however, that British Library services should be used since they were the National Agency, and were conveniently placed geographically.

It was now essential for Hatfield to move quickly because of the additional factor that the existing card catalogue production equipment was in desperate need of replacement. The County Library underwent internal reorganization to create a centralized cataloguing unit and decided to withdraw maintenance both of regional catalogues and purchase of BNB cards for the college libraries. As soon as the retrospective MARC microform database became available, Hatfield purchased four sets and commenced another major project during the summer vacation of 1975.

6
Database creation

The Hatfield library staff had become adjusted to the need for major reorganizations to provide a new dimension of library service, and once again split into teams during the summer to compare each title on the shelves with the retrospective MARC file, in the hope of recording a control number for subsequent data capture, and a Dewey classification which might prove acceptable to library users (Fig 4). Several things emerged during this exercise! Firstly it was interesting to note that microfiche files were preferable to microfilm files, even for a very large file, although this fact should have been evident from the earlier BUCCS research project at Bath University library, where various forms of library catalogue had been studied.[13] Of the four sets of MARC files, two were obtained in cassette film (12 cassettes) and two were sets of nearly 100 microfiche. By popular request, the cassettes were quickly exchanged for fiche files since these proved easier to handle in constant use. This experience also determined the output form of the eventual automated catalogue. The second factor to emerge was the poor state of the machine produced retrospective file. Many gaps were found as a consequence of the haste in which it had been prepared and many inconsistencies in BNB practice over the years were highlighted.

Experience was also to show that BNB classification decisions were not always acceptable in the local situation. The entire schedules for biology had to be rethought in a different facet order to match Hatfield teaching patterns, and often the BNB classification of new subject areas lacked sufficiently detailed expert knowledge. Such experience helped to show that classification in a user-oriented library must be a decision made by librarians in close consultation with users, although a standard scheme should be followed as closely as possible to hold down costs. Catalogue descriptions, however, can be standardized and use of MARC coding prescriptions provided cataloguers with more help than hindrance.

During this exercise, two types of catalogue worksheet were prepared, (Figs 5, 6) a CAT1 form recording only control number and local accession and classification data, together with sufficiently brief author/title data to identify the record quickly. If no MARC record could be found, cataloguers prepared a CAT 2 form, preprinted with MARC tags to record all necessary detail.

Early experiences with British Library's Local Cataloguing Service (LOCAS) were illuminating and instructive – to both sides. The City of London Polytechnic Library commenced using LOCAS before Hatfield, and found a regrettably high error rate in their output. In consequence, Hatfield decided to write programs for their DEC-10 main frame computer in order to carry out several validity checks on the data (eg 'did the string include an ISBN, BNB, or LC control number'), before input to British Library. To reduce costs, input was to be submitted by Hatfield in machine-readable form to facilitate matching against MARC tapes. However it quickly emerged that whilst the Polytechnic Computer Centre could punch cards, the mainframe was disc based with no magnetic tape facility, an input medium preferred by British Library. A further data conversion stage was required, therefore. The data-vetting stage at Hatfield proved useful, particularly in view of the continuing process of identifying stock and preparing machine-readable records. For the first 18 months, data sheets were prepared and punched, but there was insufficient data to warrant a major processing run at the British Library to produce a formatted file and a COM catalogue. Quite often, therefore, cataloguers would feel sure that a record had already been created but would be unable to retrieve an earlier worksheet from the pile of boxes of sheets! The DEC-10 run thus helped to check on whether the record had been requested before, in which case a second copy message could be recorded.

The resulting batch process proved complex and vulnerable to mishaps at several stages (Fig 7). Batches of worksheets would be punched and validated; this was done at Hatfield, or, when the volume increased, by a bureau in Birmingham which collected and delivered weekly. Card batches would be run

outside peak hours on the DEC-10, either overnight or at week-
ends but usually on a monthly cycle timed to coincide with the
British Library schedule. False starts or machine crashes due to
the large files were not uncommon resulting in regular panics
and missed deadlines. DEC-10 output could be converted into
magnetic tape by arrangement with a friendly local industrial
firm's computer, or boxes of cards would be taken by van or
train to the British Library in London. Production running,
however, was at Harlow on another bureau computer. Run
failures at this stage were by no means uncommon because of
the number of factors involved, but determining whose fault
this was presented many difficulties. When a successful run
had been made, a COM bureau at Watford became involved
with eventual delivery to Hatfield. LOCAS software at this stage
permitted only annual cumulations and monthly cumulating
supplements. A source of grievance was an extra cost involved
each year in reformatting files to produce the new cumulation
and the inability of the software to maintain a union database
which could be shared by all LOCAS users, a feature offered
by the BLCMP system.

Online cataloguing seemed the obvious development, but
remained an elusive chimaera for some time. The EDITOR software promised by BL took a long time to develop and appeared
costly in operation when finally offered. Experimental projects
were considered to compare parallel operation of batch inputting and online editing, and attempts were made to link with
OCLC and with UTLASS in Toronto. At this early stage the
Post Office reluctance to allow dataline access to commercial
telephone companies in the United States, a problem which had
bedevilled the early period of online information retrieval experiments in the STEIN project.[9] The attempt to link to UTLASS
proved almost hilarious, since separate telephone lines could be
found linking Toronto to Halifax, Halifax to London and
London to Hatfield, but attempts to send intelligible messages
failed because of technical problems.

Batch operation cataloguing using LOCAS started at Hatfield
in 1976. By May 1977 the British Library was able to offer
monthly access to a better retrospective file (back to 1950) and
the service certainly began to improve, especially as more staff
were added to provide customer support. It had been discovered that the original retrospective file had mislaid some 40,000
records, largely from the period 1950–69 but many of these had
been recovered.[14] However planning work was also in hand at

BL for a radically different approach to database organization which would provide eventually an online service. News of this MERLIN project* emerged only slowly and it depended upon the eventual acquisition by the British Library of its own ICL 2970. MERLIN incorporated a distributed database concept similar to that developed at Dortmund for DOBIS, involving a single axial control file and parallel files of common elements of catalogue entries, theoretically thereby reducing storage needs but involving complex data string creation for each catalogue entry.[15] It had been envisaged that development and implementation would take place between 1976–8, with acquisition of BL's own machine by 1978 allowing the implementation by 1981 of a broad range of online services. Continued uncertainty about future operations made it difficult for LOCAS staff to provide firm promises of improved services, or to be confident of forward cost predictions. Early thinking assumed that the British Library could support some 40 customer libraries in the first stage of development in 1978–81. At the time BLCMP was providing for only nine member libraries, and did not expect to cope with more than 20 (in 1982 it has over 30). A concept of several regional networks being developed thus seemed sensible, the alternative being that individual libraries should continue to develop separately using local authority computer services where these were available.

*[b.1975, d. 1979. Other systems with some elements of this sort of file organization include LMR's ADLIB and the ALS online circulation system. *Ed.*]

7

Issue system automation

Computerized cataloguing was well underway both for Hatfield and the Hertfordshire Colleges by 1978, but the problem of automating the issue system remained unsolved yet a high priority. In the early thinking, the issue system had been seen as the most labour-intensive task, and one with the highest priority for computer assistance. However because of the lack of an accession number listing of the stock, it seemed that a computerized issue system which could have handled accession and borrower numbers was doomed to await completion of the catalogue files. Nevertheless hope had been rekindled through meeting a small and local electronics firm, S. B. Electronic Systems Ltd, who had developed a very efficient bar code reading light pen, and successfully linked this to a microprocessor for stock handling purposes (Fig 8). The inventors, George Sims and Chris Barnett, had speculated whether their ideas could have a library application and luckily talked to their local polytechnic library. Peter Evans of the Polytechnic Library staff was given the task of considering how best the S. B. Telepen could be adapted to control the issue of books.

The Telepen and its associated Intel 8080 microprocessor provided a very interesting low cost approach to book issue control,[16] accepting that it essentially provided only a facility which sorted item and borrower numbers, and would have to be backed up by other files to identify these. Discussions centred around the fact that a polytechnic library may not have high volumes of issue transactions, perhaps only 200 – 300 per day, but it does have a complex variety. Factors such as several different borrower statuses, each with varying quantities of loans permitted, several lengths of loan categories for types of books and the continual need to trap selected items, or borrowers, are features which distinguish academic from public libraries. Hatfield had taken careful note of operational research studies at Lancaster University and had identified a core of some 10,000 volumes which were titles in heavy demand. It was thought that if these were put into short-loan categories and this part of the issue provided with computerized assistance, then considerable staff effort could be saved with additional benefits from improved turn-round of material.

The prototype Telepen stand-alone issue system was developed for Hatfield in 1977 and proved to be marketable in the £15,000 bracket – very competitive compared with other systems. Each Telepen unit with its microprocessor could hold about 5,000 live loan records, matching book and borrower numbers. Expansion was allowed for by linking units together. Although the device relied upon a volatile memory, ie if power ceased all records vanished, it proved possible to have an inbuilt battery for safety and data was also simultaneously passed to a magnetic tape cassette for historical record processing and file restoration if necessary. With the aid of a programmer in the Polytechnic Computer Centre programs were written to enable the DEC-10 main frame to generate bar-code labels on a Diablo printer-terminal. Borrower numbers were linked to the existing student records already held on the mainframe and book numbers were related to a Short Title Catalogue Listing. Preparation of this listing was largely by clerical staff on the library counter who were asked to prepare brief entries within a fixed field of about 60 characters. The list was intended only as sufficient identification of a title to produce meaningful overdue notices, or to help a reader recognize what he had on loan. (Fig 9) A second Telepen unit was purchased to act as back-up and to carry out overdue checks regularly. The system was designed to recognize four categories of borrower, each permitted different numbers of items on loan, and books in three different loan periods. A 32-character alpha-numeric display provided instant display of borrower state, indicating when necessary that borrowing limits had been reached, or that the reader was in default by keeping material too long (Fig 10).

The equipment proved simple to operate and effective in operation. Problems arose from the fact that it was prototype equipment and subject to a few mechanical faults such as poorly soldered joints! More problems were encountered, however, with the ancillary equipment and especially the tape cassette back-up unit. This was later cured by changing to more reliable 'floppy disc' drives. A particular feature of the system is the extensive use of bar-code labels for instructing the system, or for inputting sub-routines to change parameters (Fig 11).

The urgent problem of the issue system was thus partially solved with the introduction of Telepen, but for the library to achieve substantial benefit the system needed to be improved to include all library loans, to provide better enquiry facilities, to provide access to bibliographic and reader files and to give a better analysis of the use of stock. During 1979 therefore, all the library's automated systems were reappraised and a fundamental shift of development agreed. The time had come to move to online operation to improve control and to obtain a much greater degree of integration between systems. After a close study of alternatives, it seemed clear that a move to the SWALCAP operation would provide advanced facilities at a reasonable cost, and a move could be made with confidence to a proven system developed along the latest lines. A full description of SWALCAP services and their adoption at Hatfield is contained in Part 2.

8
Integrated online planning

By 1980 the Polytechnic was successfully operating a variety of independent automated systems:

1 A cataloguing system using BLAISE/LOCAS batch operations to produce two separate catalogue sequences: one for the Polytechnic Library stock and the other a union catalogue developing to include the 12 Colleges of Further Education libraries in Hertfordshire.
2 A stand-alone microprocessor-based issue system for the high-volume use material in the Polytechnic library.
3 Several in-house programmes running on the DEC-10 mainframe to produce serial listings and subject indexes.

The disadvantages of batch operation had become obvious – delays due to the variety of links in the chain of operation, and the lack of immediate knowledge of recent data input. On the other hand, a complete retrospective catalogue file had been created for the Polytechnic, and work was well underway on building up the Colleges file. A study of alternatives which could provide a more integrated system linking order procedures, cataloguing, issue control and information retrieval indicated two possible lines of development. One would be to purchase a dedicated minicomputer and appropriate software packages. The other was to move to another bureau operation offering more comprehensive services than the British Library were ready to provide.

During 1979/80 a number of package combinations had begun to appear, for instance the STATUS free text system developed at AERE Harwell was considered attractive* and GEAC had

*[See the Case Study in this series: *Studies in the application of free text package systems*. LA Publishing Ltd, 1982. *Ed.*]

appeared from Canada as a complete computer plus software
'package' designed specifically for large academic libraries. The
complete service options provided by Oriel Computer Services
were available at published rates. Of the major library co-
operatives, BLCMP was able to offer existing member libraries
the possibility of converting to online operation under BOSS,
and SWALCAP had reached the stage of offering well proven
online services to libraries outside the Southwest. Quite apart
from technical considerations one major factor was the question
of how such operations could be financed within the constraints
of local authority budgetting. A second consideration was the
extent to which purchase of dedicated equipment for the library
might require the services of specialist staff within the library
team. Direct comparison among the alternatives is not easy
because so many factors have to be taken into consideration,
but publication of a directory by the Centre for Catalogue Research offered a useful lead.[17] However study of the relevant
pages on costs for the alternative services listed there illustrates
the problem since each of the co-operatives or services had a
different basis for assessing the cost of their operation.

For Hatfield the choice quickly narrowed to SWALCAP,
partly because local policy and financial control meant that it
was easier to finance annual running costs in buying services
than to make a case for a large single capital investment, probably of the order of £100,000, to purchase a dedicated minicomputer. The appeal of the black-box approach was also most
attractive, since SWALCAP provided the computing development and operational expertise and the local library requires
only a 'computerate' librarian as the designated contact, able to
participate in joint discussions between member libraries.
SWALCAP also appeared to offer the best variety of well proven
systems designed around the needs of academic libraries yet
offering considerable freedom for the individual library to select
those features which best matched local requirements. It is interesting also to note how SWALCAP was conceived as a local
network for Southwest England, but because of its policy of
sharing telecommunication costs, the take-up by libraries far
distant from Bristol had accelerated rapidly during 1980/81.[18]

It was decided that Hatfield would move its files from LOCAS
to SWALCAP in January 1981.

It was also intended that Hatfield would implement the
SWALCAP issue system by September 1981 and this depended
on first stripping the full catalogue file to produce a short title

'bibliography file' to drive the issue system. The summer vacation at Hatfield involved a relabelling exercise to ensure that all books received a new bar code label with a number conforming to the SWALCAP system. The exercise was greatly facilitated, however, by the ability of SWALCAP to provide a printout in classified order.

Bar code labels using the SWALCAP format were printed by Telepen and supplied ready mounted on sheets, and the totally new sequence of numbers followed the SWALCAP classified print-out short-title file (Fig 9). The benefit of adopting a proven system was demonstrated when the issue system was switched on for the new session, *one day before the start of term* but it functioned immediately.

It is worth noting that Hatfield has probably been one of the first UK libraries to transfer large catalogue files from one co-operative to another. The relative ease of the operation for Hatfield demonstrated the value of buying in an outside service. At a much earlier date, Lancaster University had transferred issue files on an internal computer system as a consequence of a change of mainframe computer, but a period of near chaos had ensued.[19] Nevertheless, SWALCAP itself had been having problems due to a change of central processor and late delivery of equipment but this was not allowed to have a serious affect on customer services.

Transfer of the catalogue files from LOCAS to SWALCAP provided also an opportunity to rethink the format of entries. The original design of the LOCAS output on to COM fiche had taken into account research undertaken at that time by NRCd concerning presentation of material on a microform reader screen.[20] All too often, libraries simply transfer their previous practices which suit 5 x 3 cards or sheaf and page catalogues, but do not take into account the benefits offered by a change to microform files. For instance, since quantity of paper or numbers of pages and cards are not a problem when using microforms, it would not cost very much to present only a few entries on the screen, using large type if required. The original Hatfield catalogue file therefore spaced entries generously and averaged eight catalogue entries per frame of microfiche (Fig 12).

Later research studies were carried out at the Royal College of Art by Linda Reynolds and her report on layout of catalogue entries proved invaluable.[21] In particular she demonstrated how layout and typographical design could significantly improve readability of entries, and showed how certain items such as

location data could be perceived more easily by the catalogue
user. Recent work by the Centre for Catalogue Research at Bath
University has also concentrated the mind on length of catalogue entries and the relative value of elements of each entry.[22]
As a result the Hatfield and Hertis catalogue layouts were redesigned when drafting the SWALCAP output specification. By
attention to detail, the number of entries per frame was doubled, without sacrificing legibility and ease of use. This was
combined with a change from 42x to 48x reduction ratio and
almost halved the quantity of fiche required for the catalogue
(Figs 12 and 13). As had been noted in earlier studies,[23, 24] COM
costs become a signficiant item over even a short period of years
when full cumulations are required. Additionally, however,
SWALCAP noted that the Hatfield catalogue run required only
half the computer processing time of its other customer libraries,
thus providing another significant element of cost saving.

9

Systems review – 1979

The unsatisfactory performance of the cataloguing system led, in 1979, to a review of all automated systems. This showed a variety of independent systems, each having taken a separate line of development to meet its particular problems:

Cataloguing

The batch cataloguing system used the Polytechnic mainframe as part of the BLAISE/LOCAS service. Two separate union catalogues were produced, one for Polytechnic stock, the other for the Colleges of Further Education in Hertfordshire (HERTIS). Its achievements were substantial:

- multiple copies of the complete catalogue, with more access points and better geared to user needs
- elimination of card filing and its attendant errors
- removal of a cataloguing backlog
- transfer of professional staff from cataloguing to reader services
- catalogue records available in MARC exchange format, suitable not only for transfer to other automated systems, but also for automatic conversion to new cataloguing standards.

But there were also the problems inherent in any batch system:

- duplication of effort in the creation and input of data
- poor currency and security of data. Experience had demonstrated the need to maintain local records for data in

process, partly nullifying the economic benefits of a
computer-based operation
- the inevitable proofsheet paper-cycle
- the unsatisfactory, indirect access to catalogue records in
  progress

and those problems specific to Hatfield's operation of the system:

- the use of a local system operating on the Polytechnic
  DEC 10 machine to validate data before input to LOCAS
  and to check whether a record had been requested recently
  was not working satisfactorily. The loss of the source
  programs made it extremely difficult to correct faults and it
  was suspected that some data was being lost
- the use of an external data preparation bureau added
  another link to an already long chain stretching from data
  creation to catalogue production. An equipment failure or a
  missed deadline often meant the non-appearance of a
  catalogue.

LOCAS had not lived up to its early promise. Access to catalogue records was still limited to those produced centrally – the UK and LC MARC files. The intention for it to become an online operation had not been realized, and the link to the BLAISE Editor had not been satisfactory.

While the benefits of a computer-based cataloguing system were evident in the much improved service given to library users, the operation of the system had serious defects and its replacement was imperative.

Circulation control

The stand alone, microprocessor-based issue system had been developed using the S.B. Electronics Telepen to control the circulation of stock in high demand. The new Mk3 unit was reliable, giving online overborrowing control, reader and book trapping facilities and listing overdue books. Its main inadequacy as a stand alone system was its inability to provide directly reader or bibliographic details.

The encouraging operation of this limited system made the introduction of a similar system to control the circulation of the entire stock highly desirable, to remove many of the restrictions on the quality of service imposed by manual operation.

Several in-house systems were operating on the Polytechnic mainframe, producing subject indexes for the Polytechnic and the Colleges of Further Education in the county and a union list of periodicals in Hertfordshire.

These systems were working satisfactorily and were not candidates for change. The Union List program provided the facility to produce individual library holdings lists, and several local special libraries had accepted an invitation to include their own holdings and to receive a local list updating service.

Book order system

This was still operating manually, but was appropriate for automation both because of the inadequacies of the existing system and also for its links with Cataloguing and Circulation.

The next stage in this reappraisal was a definition of the requirements for the further development of automated systems. These were approached with Cataloguing chiefly in mind, but also with regard to Circulation Control:

Cataloguing

To maintain the current level of service:

1 It must be MARC-based with facilities for retrieval of records from UK and LC MARC files and receipt of these in MARC exchange format.
2 Addition of local data to these records and local creation of full records must be convenient and reliable.
3 The new system must be able to accept the Hatfield and HERTIS existing files, or be able to convert them to its internal format.
4 It must continue to support the production of two union catalogues.
5 Running costs should be comparable to, or less than, those for the current system.

To enhance the operation of the system:

6 The amount of original cataloguing should be reduced by access to co-operatively produced records.

7 Access to the master file of catalogue records should be online or at least via Direct Data Entry record request with overnight turnround, and VDU and printer options should be available.

8 Online editing facilities should be available, including record creation, deletion and copy. Within a record, field and sub-field editing should be possible.

9 As cost restrictions could make a completely online system unrealistic, the system should blend the advantages of online with the economy of batch operation.

10 Cost again could preclude a fully integrated acquisitions/ cataloguing/circulation system, but the cataloguing system should be able to generate short bibliographic records for use with a circulation control system.

To meet the longer term aims of the automation programme:

11 Plans should allow an even greater sharing of data as part of a national network.

12 The system should potentially be part of a fully integrated, online system, accepting data from Acquisitions as embryonic catalogue records and generating records for Circulation.

Circulation Control

1 Data capture should be at least as convenient as that used by Telepen.

2 Files should be updated in real time.

3 Online access should be provided to the Loans, Bibliographic and Reader Files.

4 Bibliographic and reader details should be added to overdue notices automatically.

5 Different reader categories should be available and the application of different loan conditions to each category should be flexible.

6 Online overborrowing control and reader interception should be provided for items in high demand.

7 A very short loan period (less than one day) should operate for items in high demand.

8 Analysis of the use of the system should be possible.

Given these requirements the options available were as follows:

1 Improve the present local catalogue control system for input to LOCAS. This would be a relatively cheap and (hopefully) early solution, but it would remain basically a batch operation with the inherent inflexibilities and data control problems. Requirements 6–12 above would not be met. This was regarded as a last resort.

2 Develop an in-house system to maintain catalogue files, passing them to LOCAS (or BLCMP) for formatting. This would meet requirements 1–10, giving tight data control, but would require large amounts of disc storage and substantial programming and data processing commitment. The use of the Polytechnic mainframe to support the academic and research needs of the institution made any substantial long term commitment to administrative systems difficult.

3 Develop a more complex in-house sytem which also performed output formatting. Although overall costs might be lower the risk could be greater.

4 Join an existing co-operative.
Requirements 1–10 were likely to be met, (11) and (12) would be possible. Other advantages were substantial:

- costs and expertise were shared
- the staff of the co-operative were likely to combine expertise in computing with a detailed knowledge of library operations
- capital costs were relatively small as the most expensive pieces of hardware are owned co-operatively (this fits the local authority approach to financial control, favouring annual revenue rather than capital expenditure)
- supplies were committed to improving existing systems and developing new ones
- systems in operation were tried and proven
- while the central machine would not be dedicated to Hatfield, at least it would be dedicated to library operations.

The main disadvantage of this alternative, was that all costs were visible and payable to an external agency, rather than 'hidden' transfers of funds between Polytechnic departments.

5 Buy a turnkey package such as GEAC.
The system would be dedicated; and likely to meet all

Circulation

The options for circulation lay between the co-operative, turnkey, and an expansion of the stand-alone system. Choice would be dependent on the decision taken for cataloguing.

The option most appropriate to local circumstances at Hatfield was membership of a co-operative. The shared approach to the improvement of existing systems and the development of new ones seemed to be the only way to keep in touch with a rapidly changing technology without investing large sums of money in hardware, which would probably be superseded in the middle future.

The choice lay between SWALCAP and BLCMP. OCLC were making tentative soundings of opinion in the UK, but their plans seemed unlikely to be revealed for some time. At that time BLCMP were operating a batch cataloguing service and had announced its intention to:

− add online features to enhance it
− add an acquisitions and ordering module
− develop a stand-alone turnkey circulation control package.

SWALCAP had been operating an online circulation control system and a cataloguing system with online features for some time. An upgraded version of the former was at an advanced stage of preparation, to be followed by a second version of the latter. An ordering and acquisitions module was also planned, completing what would become an integrated system.

SWALCAP was chosen primarily because it was operating two well proven online systems and also because of its flexible approach. Systems are designed so that each library can make its own decision on whether or not the benefits of a particular feature justify the additional cost. Although with fewer members than BLCMP, LOCAS or SCOLCAP, SWALCAP was financially stable, with membership continuing to increase.*

Having accepted that SWALCAP was the most appropriate co-operative from a systems point of view the crucial factor was then the cost of the service. Preliminary discussions with

*[The development of SWALCAP is the subject of a future Case Study in this series. *Ed.*]

38 SWALCAP led to several drafts of a document specifying the system and estimating capital and running costs.

1 A comparison of the likely running costs for the first full year's operation of the SWALCAP Cataloguing system with those for the BLAISE/LOCAS service in the 80/81 financial year gave the following figures:

	LOCAS	SWALCAP
Record selection, processing and formatting	£14,500	£16,000
External data preparation of EMMA	£8,000	–
COM production	£8,500	£ 8,000
	£31,000	£24,000

These figures are for the Polytechnic and HERTIS catalogues. The LOCAS figure is for an annual cumulation of the catalogue with monthly cumulating supplements, the SWALCAP figure for complete bi-monthly cumulations. Data preparation is excluded from SWALCAP charges, as with online operation it would be done internally using existing staff resources.
SWALCAP costs were regarded as more likely to remain stable or even be reduced, because of an increasing membership to share them, while LOCAS charges were rising steadily.
Circulation running costs for a full year's operation were estimated at £11,000.

2 Non-recurrent costs for equipment, catalogue reformatting, systems development and AACR2 conversion were estimated at £34,000. General recurrent costs including equipment maintenance, network charges and modem rental were likely to be £4,000.

The capital costs could be met in the 1980/1 financial year, while recurrent costs were acceptable, especially so with the prospect of further economies of scale resulting from a growth in membership. In January 1982 SWALCAP announced for the third successive year that its processing charges would not be increased and in February 1982 COM production costs were reduced when the volume of SWALCAP fiche production reached a level justifying a lower tariff.

10
Systems description

SWALCAP is a non-profit making organization based at the University of Bristol. At the time of writing it has 17 members (universities, polytechnics and the BBC Reference Library) subscribing to one or both of the systems currently available. Most of the members are located in the South-west and Wales, although membership is extending rapidly throughout the South-east, Midlands and further north.

Cataloguing

The SWALCAP Cataloguing System provides facilities to create and maintain a catalogue in machine readable form, together with its production on COM fiche. All activities concerned with the creation or amendment of catalogue records are carried out online at a computer terminal – usually a visual display unit – which is linked via a minicomputer at Hatfield to the main SWALCAP computer. The minicomputer validates control numbers and holds catalogue records in current use. Cataloguing can operate if the link to the central computer is broken (Fig 14).

The catalogue records for each user are held on two machine-readable files:

– the Catalogue File, which holds the complete catalogue fully formatted, ready for its next production on fiche. As this file is too large to be held online for the user to access directly, any catalogue records which need amendment, including the addition of extra copies, are first transferred to –

– the Supplement File, a much smaller working file (or catalogue-in-progress file), which is held online and thus directly accessible to the user at a terminal. All records of current interest are held here – not only those recalled from the catalogue for amendment but also recently acquired records awaiting checking and editing before release to the Catalogue file and inclusion in the next fiche catalogue.

New records are added to the catalogue in several ways:

- from the catalogue files of other SWALCAP members
- by purchase from external organizations such as the British Library
- by creating them locally from scratch.

A catalogue record – whether an existing one on the user's Catalogue File which needs amendment or a new one for an item recently added to stock – is requested by keying in the control number (ie ISBN, BNB, LC, SWALCAP No.) at a terminal. The user's own Catalogue File is first searched overnight. Where the request is for a new record the search then continues through the files of other SWALCAP members again on an overnight basis and finally, if still unsuccessful, to external sources, BLCMP in particular. Any records located in the user's own files, or those of other SWALCAP members are transferred to the requesting user's Supplement File, available for editing the next day. Requests sent to BLCMP are matched against its files on a weekly basis.

Any record held on the Supplement File can be called to a VDU screen by keying in its control number. Once retrieved its contents can be manipulated using a set of commands which are executed as part of a simple dialogue between machine and operator. Single commands such as INSERT, ERASE and CHANGE can be used to alter single letters, words or even complete MARC fields. Amendments are applied immediately and are displayed on the screen for checking. When all work on a record is completed it is transferred through the Supplement File back to the Catalogue File to enter the next catalogue production cycle.

Where no control number is available to initiate a search for a record, or where a requested record has not been obtained within a specified time, a catalogue record is created from scratch at a terminal. After checking it is transferred to the Catalogue File for inclusion in the next fiche catalogue.

The system controls the displacement of all library stock, whether borrowed by readers or removed for any other reason such as inter-library loan. It operates online and files are updated in real time – when a transaction is executed the relevant files are updated immediately and subsequent consultation of those files will give a prompt and accurate response.

Issue, return and renew transactions are performed at Hatfield by scanning reader and book bar-coded labels with a fibre optic wand manufactured by S.B. Electronics (the Telepen). The Telepen is used only as a means of data capture and display, all validation and processing being done by the on-site minicomputer or the central computer situated in Bristol. The minicomputer controls the terminals, validates each transaction before sending it to Bristol and holds the Interception List of trapped users, and password details. The central machine performs further validation of the data, updates the loans file and sends back responses to the relevant Telepen unit.

While cataloguing is dependent on the link to the central computer, circulation can still be operated without it, although facilities are limited to issue and return. No file enquiries are possible as they are held centrally. When off-line, transactions are recorded to cassette and then transmitted 'down the line' to update files when the link is restored.

A wide range of facilities are available:

1 VDUs situated at the library counter (or even those used for cataloguing) enable online interrogation of a number of files:
   - the Loans File to ascertain if an item is on loan, who has borrowed it and when it is due back
   - the Bibliography File (short author/title records). Enquiries by author/title, classmark or book number retrieve a record containing both bibliographic details and loan status
   - the Registration File, containing reader details and number of books on loan. Access is by reader name or number.
2 Overdue and reservation notices are printed locally on a daily basis. Reader and bibliographic details are added automatically.
3 A varied loans policy can be operated to control any combination of user category, loan periods and material

types. The conditions governing this matrix can be easily amended.

4 A wide range of statistics on the use of the system can be produced on demand. When an item is returned from loan a record is created and archived. Each record contains a comprehensive list of loan details:

Book number	Loan status
Reader number	Loan period
Loan date	Days on loan
Loan time	Days overdue
Return date	Notices
Return time	Reservations
	Classmark

These details can be combined to produce an analysis of loans. By linking them with bibliographic details a more detailed analysis of individual titles and copies is available.

5 A by-product of the system is the use of the Bibliography Files to produce selective listings of items in stock.

11
Systems implementation – 1981

The preparatory work for both SWALCAP systems was conducted throughout the first nine months of 1981. Cataloguing was scheduled to go live in March and circulation in September.

Cataloguing

The transfer of catalogue records from LOCAS to SWALCAP and their induction to the SWALCAP system involved a number of steps:

1 *Transfer of files*

Both the Polytechnic file (c.110,000 records) and that for HERTIS (c. 70,000 records) were written to tape in MARC Exchange format by LOCAS for transfer to Bristol. Even this relatively simple exercise had to be repeated several times because of poor quality tapes.

2 *Conversion to AACR2/MARC2 format*

The timing of the files transfer coincided with the introduction by the British Library of AACR2 and MARC2. When LOCAS announced its intention to convert its members' files to the new formats it was hoped that the Hatfield and HERTIS files could benefit from the conversion before being transferred. However for continuing customer libraries, conversion costs would be written off over a period of years. Hatfield as a departing customer would have to pay the full charge immediately. While SWALCAP did not have the facilities to do the conversion, agreement was reached with BLCMP to apply its own conver-

44   sion routines to SWALCAP files. However, a series of problems encounted by BLCMP meant that no SWALCAP files had been processed by June. As the Polytechnic file was needed by SWALCAP in order to produce a shelf-list for a stock relabelling exercise (see below), the conversion plans were abandoned. The Hertis file was converted by BLCMP in mid-August, but the low number of records affected by the process (6.8%) was disappointing.

3 *Conversion to internal SWALCAP format*

The MARC Exchange format used for the files transfer ensured that the general bibliographic data in the records could be accepted by SWALCAP without the need for any special conversion, but the local data was structured quite differently and had to undergo substantial conversion to the SWALCAP internal format. The aim of the process was to construct new accession and classmark fields in the SWALCAP record from information held in different areas of the LOCAS record. It was also necessary to allocate new SWALCAP 'item' numbers to records on the Hatfield file for use with the Circulation Control system. An added bonus from this restructuring was that SWALCAP was willing to extend its range to upgrade parts of the LOCAS record so that full advantage could be taken of SWALCAP output features. Examples of the exercise included:

– the variant abbreviations for shelving locations were standardized
– a filing suffix derived from the heading fields was added to the classmark
– information in the collection and notes fields was restructured to match the redesigned output parameters
– separate reference records were created from the 900 – 945 fields.

The conversion proved to be largely successful, although the varying provenance of records and previous changes in internal practice caused a small amount of data to be lost. Some records were rejected by the process, chiefly because they were too long. All of the data lost has been retrieved and is being reinput.

The first SWALCAP catalogue appeared in October 1981, ten months after the last produced by LOCAS. This delay coincided with the late connection of the landline to Bristol and the discovery of a continuing fault in one piece of equipment, which meant that online operation of the system did not begin until mid-June. As an interim measure to reduce a growing backlog of cataloguing awaiting processing and to bridge the gap between catalogues the system was used as a batch operation between January and June 1981. Record requests were submitted with local details on punched cards and EMMA records were input at SWALCAP headquarters. A catalogue of these items was produced to supplement the last LOCAS catalogue.

Circulation

The main tasks were to create reader and bibliographic records and to prepare the bookstock.

1 Stock relabelling exercise – Summer 1981

It was necessary to insert new bar-coded labels in all items of stock before the introduction of the system at the beginning of the new academic session. Some books already contained the bar-coded labels used with the stand-alone Telepen, but these were incompatible with SWALCAP. The new 'item' numbers on the bar-coded labels were allocated automatically by assigning each accession number in the catalogue a specific 'item' number. The relabelling exercise entailed matching an accession number with its corresponding item number and sticking the relevant label in the book.

SWALCAP produced a list of all items of library stock on the catalogue in classified order. For each copy of a title the entry on the list contained accession number and new item number, plus author/title/edition/classmark details. By allocating the first entry on the list item number 1, the second 2, etc the ascending order of the item numbers corresponded to the classified order of the list and to a large extent the shelf order of books. The bar-coded labels were produced in ascending item number order so in theory at least, the first label should have been for the first book on the shelves, and so on. The exercise was complicated, of course, by books on loan or lost (label available but no book), books not in the catalogue (book available, but no label) and books mis-shelved.

A by-product of the exercise was a further improvement in the accuracy and comprehensiveness of the catalogue, as books omitted from it could be identified and discrepancies between catalogue and books corrected. Doubts about data lost somewhere in the complex batch operated LOCAS system were confirmed by the large numbers of books added to stock since 1978, which had not appeared in the catalogue.

The library was closed to the public for four weeks and at the end of that period almost all items had been relabelled.

2 *Production and editing of the Bibliography File*

The Bibliography File was derived automatically from the records in the Catalogue File, using SWALCAP's standard routine. As expected from an operation of this kind many records were created which needed manual amendment. This editing process is being tackled as an on-going exercise, concurrently with new cataloguing.

3 *Creation of reader records*

Records for full and part-time students were already held in machine–readable form by the Polytechnic Computer Centre. These were written to tape and sent to SWALCAP to form the basis of the reader file. Records for new students, staff and other reader categories were created online by library staff.

12
Systems review – 1982

As the move to SWALCAP meant that requirements for such things as catalogue output and loan conditions had to be redefined it was seen as a good opportunity to take a fresh look at all aspects of the cataloguing and circulation systems and to redesign them to take advantage of SWALCAP's online features. The design of the circulation system was regarded as more crucial because of its immediate impact on the library user. Cataloguing had undergone its most radical change when fiche replaced card as the catalogue medium. The transfer to SWALCAP was therefore largely transparent to users as its most important effect was to upgrade the operation of the system rather than to change the end product.

Cataloguing

The Polytechnic cataloguing operation takes full advantage of the online facilities available. Two cataloguers have their own personal VDUs on their desks, whilst a third VDU is situated to give convenient access for other cataloguing and clerical staff. *Catalogue records are created and amended online, direct from source to screen, without any intermediate paper records.* Combining the creation of data with its input has removed much of the duplication of effort characteristic of a batch operation and the need for typist speed input. Cataloguers have adapted readily to the keyboard and the speed of their input has quickened considerably.

The online features of the system have not only provided a more efficient and convenient means of communication, but have had a considerable effect on the organization of cataloguing routines:

- entry of data has become an integral part of the cataloguing process, rather than a separate, clerical operation. The VDU has become just another desk-top cataloguing tool, the keyboard replacing the pen, the screen the paper
- the proof-sheet paper-cycle has been removed
- the division between professional and clerical tasks has altered. Clerical input to the system includes record requests (containing local data) and simple amendments. Professional tasks are limited to classification, editing acquired records and original cataloguing. The creation of a separate category of machine operator has been avoided
- the link between the classification and cataloguing of an item has been broken; whereas both the classification and cataloguing of an item used to be done by the relevant subject cataloguer, at present after classification, record requests are done by clerical staff; editing of the acquired records is shared between cataloguers but not on a subject basis. The individual skills of cataloguers have also affected the allocation of work, as most original cataloguing is done by one cataloguer who is a proficient typist
- the short time between the initiation of a record request and its cancellation if unsuccessful (three weeks) means that books are on the shelves and a full catalogue record in existence within four weeks of arrival in the library.

In its present form the system has several limitations:

- the link with the circulation control system is unsatisfactory. Although Bibliography File entries are generated automatically when a record enters a member's catalogue for the first time, any subsequent amendment to that record is not applied automatically to the relevant Bibliography File entries
- the only realistic access to externally produced records is by standard control numbers. Any records created by SWALCAP or BLCMP members using a local control number are therefore inaccessible
- while the delay between ordering a record and obtaining it can be reduced to one day, the cataloguing process is still two-step. Even applying simple amendments to a record already held in the catalogue is laborious.

A further problem with the system was its unreliable operation in 1981, when the central processing unit was unavailable for considerable periods of time. Late delivery and commissioning of a new VAX 11/780 processor meant that the existing machine, a Xerox 750 had to support all of the cataloguing systems and most of the circulation systems longer than expected and overloading occurred. The installation of the VAX at the end of 1981 and the transfer to it of several circulation systems had since considerably improved the reliability of cataloguing.

Attention was also given to the format of entries on the fiche catalogue. The layout of entries was redesigned, guided by the results of research carried out by Linda Reynolds of the Royal College of Art (Figs 12, 13). The guiding principle was to make the most important parts of the entry clearly distinguishable while at the same time ensuring that it holds together as a cohesive unit. The most important changes were:

1 The reduction of space between entries. Large areas of space on a page are not needed to separate entries and may even make the scanning of entries less efficient.
2 The increased use of typographic and spatial coding. The usual techniques of capitalization and indentation have been supplemented with bold typeface and italics. All elements in an entry are ranged left as those placed at the top or bottom right-hand corner can become confused with the entries below and above.

These changes have improved the catalogue considerably, although there is still room for improvement. Bold typeface and italics are effective although less so on fiche than on paper, while limitations in SWALCAP formatting facilities mean that some desirable features are not available, eg capitalization of first word of title heading, rather than the complete subfield; separation of shelving location from classmark in the classified sequence; and lack of flexibility with indentation. Perhaps the most unsatisfactory aspect of catalogue entries is the need to include both item numbers and control number, neither of much interest to the catalogue user. The control number has been hidden to an extent by placing it after the imprint, while locations and item numbers have been signified with an asterisk.

Changes to the content of entries have been somewhat more cautious. The full range of access points has been retained, while the descriptive part of entries has been reduced by cutting out some elements almost completely (place of publication, collation) and using others (especially notes) very selectively. Research being conducted by the Centre for Catalogue Research at Bath may show that a further reduction is possible.

The main effect of the redesigned layout and content changes has been to reduce the size (and therefore the cost) of the catalogue considerably. The author sequences of the last LOCAS catalogue produced 150 fiche but they occupy only 60 in the new format.

Most of the limitations of the current system are likely to be removed when an upgraded version is introduced towards the end of 1982. Features under discussion include:

- a closer relationship between cataloguing and circulation control. Amendments to catalogue data will be applied automatically to the relevant Bibliography File records
- improved access to catalogue data. An index to all SWALCAP catalogue records will be held online, accessible by item numbers, author/title acronym keys and control number. This interactive facility should be particularly useful to identify and order records created with local control numbers
- the option to hold full catalogue data online. The falling costs of storage should make it economical for some libraries to hold all, or a proportion of their catalogue records online, accessible to all SWALCAP members via the online index. This will provide the opportunity to complete the cataloguing process from record order to addition to catalogue at one sitting. The number of records available in this way could be increased by holding part of the UK MARC files online
- improvements in catalogue output. A more flexible approach to output formatting will allow a much more precise definition of entry layout.

Circulation

A working party representing all areas of library activity was given responsibility for the design of the new system. A preliminary set of proposals was submitted to all library staff for comment, followed by a revised report which was accepted as the basis of the new loans policy. This will be reviewed throughout the first year of operation to allow any fine tuning of the system.

The new policy was implemented at the beginning of the 1981/2 academic session to coincide with the introduction of the new system. It was hoped that the close identification of the new policy with the new system would reduce initial resistance to the major changes in the conditions of loan.

Publicity for the system was extensive, emphasizing how much easier it would be to borrow books, the improvement in availability of bookstock resulting from the shorter loan period and the efficiency with which books on loan would be located. The aim was to make potential users aware of the much better service that the library would be supplying.

Plans for a comprehensive training programme for library staff were seriously affected by the unavailability of the sytem until several days before the beginning of the new academic year.

Three loan periods are used: four weeks, one week and four hours. The reduction of the general loan period from one term to four weeks for all readers is the most radical change; four weeks was chosen as a compromise combining an adequate length of loan with a marked improvement in the circulation of stock. Seven categories of reader have been allocated, chosen either because a separate set of loan conditions is necessary, or just to isolate them for statistical purposes. Only two categories of material are relevant at present – monographs and audio-visual material. Loan conditions such as number and frequency of overdues vary considerably with loan period but only slightly with reader category. A strict regime of prompt and frequent overdue notices is operated.

While it is perhaps too early to know the full long term effects of the new system, some are already apparent:

- the improved circulation of stock resulting from the shorter loan period has increased the amount of shelving necessary
- the removal of file maintenance work has been partly offset by an increased demand on the service, *caused particularly by the availability of new features such as file enquiry*
- the maintenance of a much tighter control over the system than was possible previously has generated extra work, eg daily production of overdues.

Although the total amount of work necessary to support the circulation system has certainly decreased, the opportunity to

52 provide a service of much improved quality and range makes
    it unlikely that any substantial reduction in staff levels is likely.

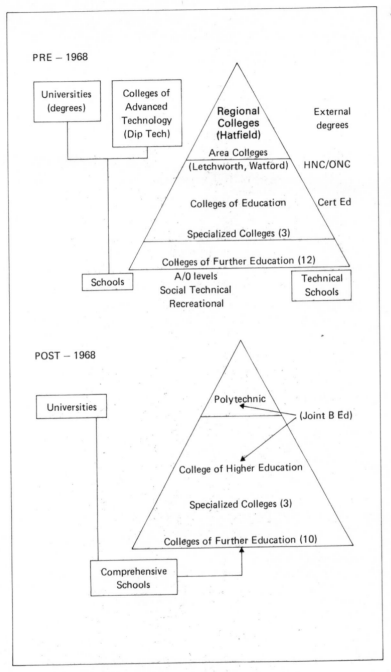

Fig 1 Further and higher education in Hertfordshire

HERTIS STRUCTURE

Function:
The basic functions of the Hertfordshire County Council Technical Library and Information Service are:

1 To develop the resources (books, pamphlets, periodicals, records, tapes, slides, films and other media) necessary to meet the requirements of further and higher education within the County.

2 To co-ordinate these resources with all other libraries within the Region

3 To provide a central agency which can use sophisticated mechanized techniques for cataloguing, information processing and retrieval.

4 To train students to use these resources effectively.

5 To co-operate with lecturers to develop their libraries as learning resource centres.

6 To co-operate with teachers through the School Library Service in order to exploit these resources for the benefit of fifth and sixth form students.

7 To exploit the resources for industry and professional personnel in the Region.

8 To develop effective links with both the national and international network of libraries and information services.

Fig 2 HERTIS terms of reference

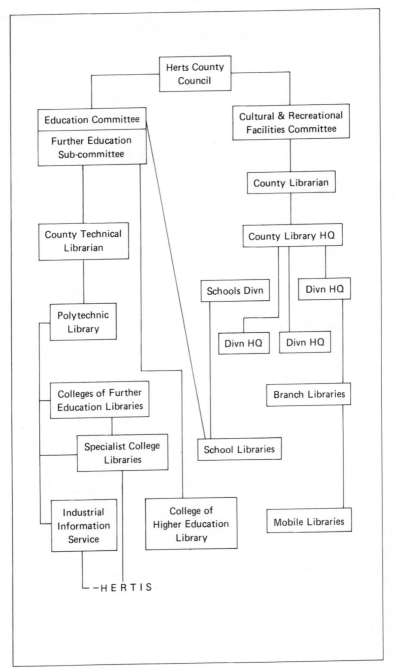

Fig 3 Library relationships in Hertfordshire

b6612384   338.911724041
   BRITISH COUNCIL OF CHURCHES (and) CONFERENCE OF BRITISH
   MISSIONARY SOCIETIES. World poverty and British
   responsibility. Published for the British Council of
   Churches and the Conference of British Missionary
   Societies by S.C.M.Press, 1966.

b6013515   270.82
   BRITISH COUNCIL OF CHURCHES. First ecumenical work-book.
   British Council of Churches, 1960.

b5306260   394.4
   BRITISH COUNCIL OF CHURCHES. A form of divine service for
   use at united services and on other occasions at the time
   of Her Majesty's Coronation. Oxford U.P.; Cambridge U.P.;
   Eyre & Spottiswoode, 1953.

0851690025   614.862
   BRITISH COUNCIL OF CHURCHES. Go for road safety. Council of
   Churches, 1968.

b5605890   270.8206241
   BRITISH COUNCIL OF CHURCHES. Going forward. British Council
   of Churches, 1956.

0851690181   200.7
   BRITISH COUNCIL OF CHURCHES. Guide to project work in
   religious studies. British Council of Churches, (1973).

b6300014   241.66
   BRITISH COUNCIL OF CHURCHES. Human reproduction. British
   Council of Churches, 1962.

0851690246   261.83420942
   BRITISH COUNCIL OF CHURCHES. Marriage, divorce and the
   Church. British Council of Churches, 1973.

b6504485   207.12
   BRITISH COUNCIL OF CHURCHES (and) NATIONAL UNION OF
   TEACHERS. Some aspects of religious education in secondary
   schools. British Council of Churches; National Union of
   Teachers, 1965.

0715155059   259
   BRITISH COUNCIL OF CHURCHES. Overseas students and the
   Churches. Published for the British Council of Churches by
   the Church Information Office, 1970.

Fig 4 LASER/MARC retrospective file

0851690378   254.
BRITISH COUNCIL OF CHURCHES. Community and Race Relations
Unit. Working Party on the Use of Church Properties for
Community Activities in Multi-Racial Areas. The community
orientation of the Church. British Council of Churches,
1974.

0851690319   254.
BRITISH COUNCIL OF CHURCHES. Community and Race Relations
Unit. Working Party on the Use of Church Properties for
Community Activities in Multi-Racial Areas. Interim repor
of the British Council of Churches' Working Party on the
Use of Church Properties for Community Activities in
Multi-Racial Areas. British Council of Churches, 1972.

b6509653   262.
BRITISH COUNCIL OF CHURCHES. Consultative Committee on
Training for the Ministry. The shape of the ministry.
British Council of Churches, 1965.

0851690238   327.420689
BRITISH COUNCIL OF CHURCHES. Department of International
Affairs. Britain and Rhodesia now. British Council of
Churches. Conference of British Missionary Societies,
1972.

0851690300   261.8
BRITISH COUNCIL OF CHURCHES. Department of International
Affairs (and) CONFERENCE OF BRITISH MISSIONARY SOCIETIES.
Investment in Southern Africa. British Council of
Churches. Conference of British Missionary Societies,
1973.

0851690351   301.324096
BRITISH COUNCIL OF CHURCHES. Department of International
Affairs. Emigration to Southern Africa. British Council o
Churches. Conference of British Missionary Societies,
1973.

0851690343   320.9512490
BRITISH COUNCIL OF CHURCHES. Department of International
Affairs. The future of Taiwan. British Council of
Churches. Conference of British Missionary Societies,
1972.

HATFIELD Cat. 1 Input

Date

Control No. | C | 1 | | | | | | | | | | | | UK 50- 74 | A
Access. Nos.

UK 50- 74	A
UK 75-	D
US 68-74	E
US 75-	F

Control No | C | 2 | | | | | | | | |
D.C. 18

Author / Title

| | Ed | Publisher | | Date |

| Source of Control No. | | | Further Action: | by: |

| Hat | BL | BIE | |
| BNB | CIP | VTP | |

Fig 5 Cataloguing worksheet – MARC material

Crl | | | | | | | | | | | | #|0|2|1| | | | | | | | | | |

008 $a| | | | | | | | | | $b| | $d| $e| $f| | | | $g| $l| $m| $n| | | $q|

009 $a| | $b| #|0|3|7| | | | | | #|0|4|3| | | | | | | #|0|4|6| | | | #|

Author 1

Title 24

Edition 250

Imprint 260 $b

Collation 30

Series 4

Notes 5

Subject headings 6

Added entries 7

Item/Accession numbers 96

Class number 970

Cat 2

Fig 6 Cataloguing worksheet, Extra MARC material (local input)

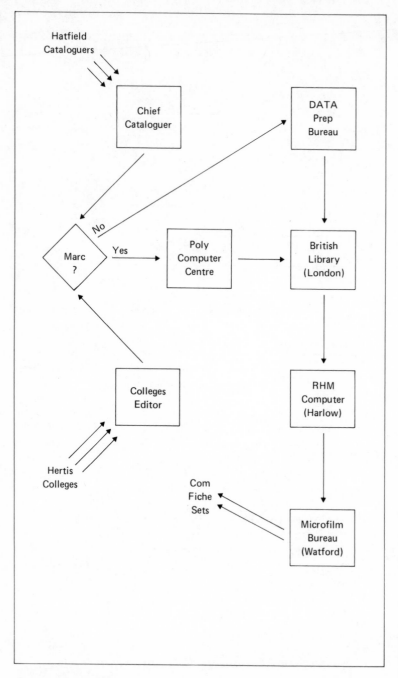

Fig 7 **LOCAS** catalogue production

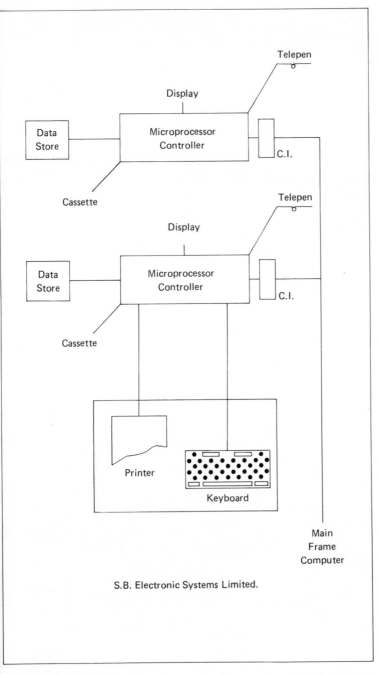

Fig 8 Telepen system arrangement

10-MAR-82 18:06   SWALCAP CIRCULATION SYSTEM ITEM FILE LISTING

UNDERGRADUATE LIBRARY COLLECTION

DESCRIPTION	CLASSMARK	ITEM NUMBER
WIGHTMAN, W.P.D. * SCIENCE IN A RENAISSANCE SOCIETY. 1972	505.09024	10 0135941 8
WIGHTMAN, W.P.D. * SCIENCE IN A RENAISSANCE SOCIETY. 1972	505.09024	10 0135939 2
WILDLIFE & COUNTRYSIDE ACT,1981 (GB.). 1981	346.4104695	10 0144446 3
WILKINS, D.A. * NOTIONAL SYLLABUSES. 1976	375.4	10 0036161 1
WILKINSON, L.P. * ROMAN EXPERIENCE. 1974	937	10 0136163 6
WILLEY, B. * SEVENTEENTH CENTURY BACKGROUND.	001.20941	10 0000009 2
WILLIAMS, B. * PROBLEMS OF THE SELF. 1973	126	10 0005505 X
WILLIAMS, G. * LEARNING THE LAW. 10E. 1978	340.07	10 0138977 9
WILLIS, P.E. * LEARNING TO LABOUR. 1977	301.442	10 0018096 1
WILSON, D.A. * POLITICS IN THAILAND.	320.959304	10 0020949 3
WILSON, E. * TO THE FINLAND STATION. 1960	335.009	10 0026072 9
WILSON, E.O. * SOCIOBIOLOGY. 1975	156.252	10 0010073 X
WINCH, R. * ENERGY,ECOLOGY,& THE ENVIRONMENT. 1974	333.7	10 0025761 2
WINCH, P. * STUDIES IN THE PHILOSOPHY OF WITTGENSTEIN. 1969	193	10 0012276 2
WINER, B.J. * STATISTICAL PRINCIPLES IN EXPERIMENTAL DESIGN. 2E. 1971	519.53	10 0045408 X
WINT, G. * BRITISH IN ASIA.	954.03	10 0099224 0
WISEMAN, S. * INTELLIGENCE & ABILITY. 1967	153.9	10 0008301 4
WITTGENSTEIN, L. * BLUE AND BROWN BOOKS. 1969	149.94	10 0136756 8
WITTGENSTEIN, L. * LECTURES & CONVERSATIONS ON AESTHETICS,PSYCHO.	193	10 0012282 4
WITTGENSTEIN, L. * NOTEBOOKS,1914 1916. 1961	149.94	10 0006119 X
WITTGENSTEIN, L. * PHILOSOPHICAL INVESTIGATIONS. 2E. 1958	149.94	10 0006117 1
WITTGENSTEIN, L. * TRACTATUS LOGICO-PHILOSOPHICUS. 1961	149.94	10 0006114 4
WITTGENSTEIN, L. * UBER GEWISSHEIT. 1969	121	10 0005398 X
WITTGENSTEIN, L. * ZETTEL.	193	10 0012292 2
WITTGENSTEIN WORKBOOK. 1970	193	10 0012284 2

Fig 9 Short-title catalogue

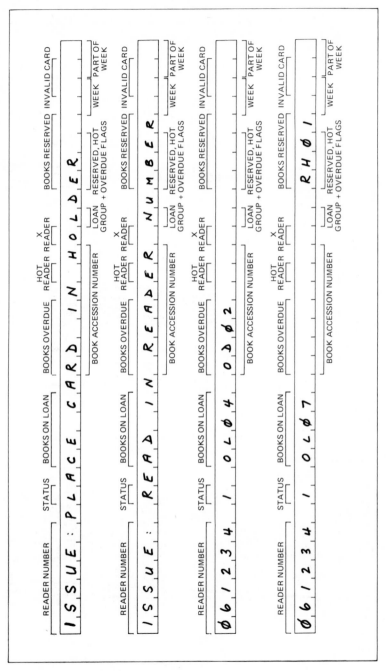

Fig 10 Telepen displays

64

TELEBOARD:2

SUPERVISORY FUNCTIONS

local   start   link

ISSUE

ENQUIRY

RETURN

RENEW

Flags

Set
Reserve book

Hot book

'X' Reader

Hot Reader

Clear

Display Reader

Display Books   First

Next   Previous

S.B. Electronics Systems Limited

Fig 11 Telepen supervisory function bar-codes

330

CARSON, Richard L
Comparative economic systems. New York: Macmillan; London:
Collier-Macmillan, 1973.
xiii, 717p. 002319510x

HB6069254 HB6069255

301.1

CARSON, Robert C
Contemporary topics in social psychology, editors, John W.
Thibaut, Janet T. Spence, Robert C. Carson ; contributors,
Jack W. Brehm ... (et al.). Morristown, N.J: General
Learning Press, c1976.
xii, 477 p. lc75040856

HT7606393 HT7719122 HT7719123
HB7719120 HB7719121

301.31

CARSON, Sean McBirney
Environmental studies: the construction of an A-level
syllabus, compiled by S. McB. Carson. New ed. Windsor:
NFER, 1973.
147p. 0856330299
Previous ed. 1971.

HT6418298

364.36

CARSON, Wesley George
Crime and delinquency in Britain: sociological readings.
London (17 Quick St., N1 8HL): Martin Robertson and Co. Ltd,
1971.
3-258p. 0855200022
Pbk. £1.15. sbn 85520-003-0.

HT5555272 HT5555274 HT5555275 HT5555271r
HB5210712 HB5555273 HB7736197 HB7736198 HB7751016

Fig 12 Catalogue layout by LOCAS

301.141

CARSTAIRS, George Morrison
      This island now. Hogarth P, 1963.
      103p. ill.,23cm. b6307985
      (Reith lectures;1962)

   HB4842742

320.53309436

CARSTEN, Francis Ludwig
      Fascist movements in Austria: from Schonerer to Hitler.
      London (etc.): Sage Publications, 1977.
      356p. 0803999925
      (Sage studies in 20th century history; vol.7)

   HT7711493

322.50943

CARSTEN, Francis Ludwig
      The Reichswehr and politics, 1918 to 1933. London: O.U.P.
      1966.
      427p. h000200484

   HT7716571

322.420943

CARSTEN, Francis Ludwig
      Revolution in Central Europe, 1918-1919. London: Maurice
      Temple Smith Ltd, 1972.
      360,(8)p. 0851170153

   HB6066720

320.533

CARSTEN, Francis Ludwig
      The rise of fascism, Batsford, 1967.
      256p.,23cm. b6719975

   HT3766746 HT7609020 HT7709281

HATFIELD - AUTHOR & TITLE CATALOGUE JUNE 1982

BUCKLEY, Henry Burton, Baron Wrenbury
Buckley on the Companies Acts. 2nd (cumulative)
supplement. - 13th ed.
Butterworths, 1968. B68-03428
*Balls Park (11 0020169 X)
Shelved at **346.41066 B**

BUCKLEY, Henry Burton, Baron Wrenbury
The Companies Acts. - 13th ed / by J.B. Lindon assisted by G.
Brian Parker and Hugh R. Williams.
Butterworth, 1957. B57-17546
*Balls Park (11 0020170 6)
Shelved at **346.41066 B**

BUCKLEY, James
Natural gas in the UK energy market.
Cambridge Information & Research Services, 1979. 0-905332-
06-7
*Hatfield (10 0025935 5)
Shelved at **0/SIZE 333.820941 B**

BUCKLEY, Jerome Hamilton
Season of youth: The bildungsroman from Dickens to Golding.
Harvard V.P., 1974. 0-674-79640-3
*Hatfield (10 0083482 4)
Shelved at **823.111 B**

BUCKLEY, Jerome Hamilton
Tennyson: the growth of a poet.
Harvard V.P., 1960. x-44-022161-6
*Hatfield (10 0087952 6)
Shelved at **841.08109353 B**

BUCKLEY, Jerome Hamilton
The triumph of time: a study of the Victorian concepts of
time, history, progress and decadence.
O.U.P., 1967. x-44-005980-0
*Hatfield (10 0005127 2)
Shelved at **115 B**

BUCKLEY, Jerome Hamilton
The Victorian temper: a study in literary culture. - 1st
ed.new impression.
Cass, 1966. B66-15228
Originally published (B52-2857) Allen & Unwin,1952,
classified at 820.9.
*Hatfield (10 0086933 1)
Shelved at **841.081 B**

BUCKLEY, John W
The accounting profession.
Melville Pub. Co, [1974]. LC74-008880 (Melville series on
management, accounting, and information systems)
*Balls Park (11 0028686 4)
Shelved at **657.023 B**

Fig 13 Catalogue layout by SWALCAP

BUCKLEY, John W
  Contemporary accounting and its environment / [edited by]
  John W. Buckley.
  Dickenson, 1969. LC74-078800 (Dickenson Series on
  Contemporary Thought in Accounting)
  *Balls Park (11 0051149 4)
  Shelved at **657 B**

BUCKLEY, John W
  Research methodology & business decisions.
  NAA, c1976. LC76-001393 (NAA publication ; no. 7581)
  *Balls Park (11 0030957 1)
  Shelved at **658.403 B**

BUCKLEY, Laurence
  Buying and selling your home.
  Oyez PublishingWard Lock, Oct. 1979. 0-7063-5821-x
  *Hatfield (10 0025625 0)
  Shelved at **Q/REF MF 333.337 B**

BUCKLEY, Marlene H, b.1938
  The accounting profession.
  Melville Pub. Co, [1974]. LC74-008880 (Melville series on
  management, accounting, and information systems)
  *Balls Park (11 0028686 4)
  Shelved at **657.023 B**

BUCKLEY, Marlene H, b.1938
  Research methodology & business decisions.
  NAA, c1976. LC76-001393 (NAA publication ; no. 7581)
  *Balls Park (11 0030957 1)
  Shelved at **658.403 B**

BUCKLEY, Peter
  The Spanish plateau: the challenge of a dry land / maps by
  Wesley McKeown.
  Chatto & Windus, 1962. B63-09202
  For children.Originally published.Coward-McCann,1959.
  *Hatfield (10 0093476 7)
  Shelved at **914.6 B**

BUCKLEY, Peter Jennings
  The future of the multinational enterprise / [by] Peter J.
  Buckley and Mark Casson.
  Macmillan, 1976. 0-333-19476-4
  *Balls Park (11 0015690 1)
  Shelved at **338.88 B**

BUCKLEY, Peter Jennings
  Going international: the experience of smaller companies
  overseas / [by] Gerald D. Newbould, Peter J. Buckley, Jane C.
  Thurwell.
  Associated Business Press, 1978. 0-85227-205-7
  *Balls Park (11 0030107 7)
  Shelved at **658.18 N**

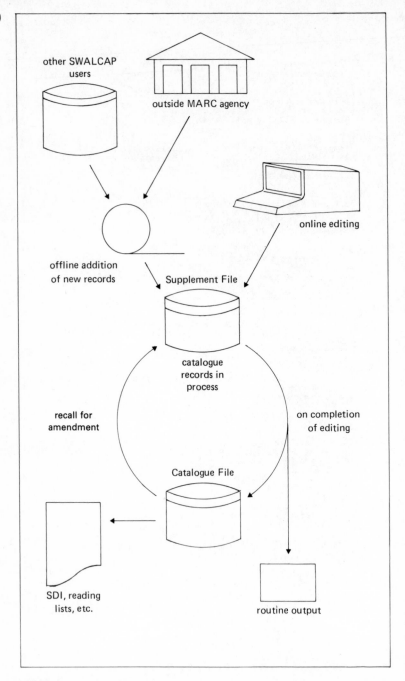

Fig 14 SWALCAP cataloguing system

1952–66	Creation and development of HERTIS network of libraries supporting Further and Higher Education. 71
1960	Introduction of centralized catalogue card preparation using embossed metal plate equipment.
1962–7	Investigations of mechanized microform retrieval and creation of National Reprographic Centre for documentation.
1967	Purchase of paper-tape typewriter to prepare machine-readable data.
1968	In-house creation of subject index by computer.
1968	Designation of Hatfield College as a Polytechnic followed by considerable expansion.
1970	Physical move of main library to four-storey building with second site requiring library services.
1970–2	Negotiations with English Electric Ltd and Oriel Computer Services for catalogue production.
1971	Complete stock check at Hatfield slims catalogue to c50,000 live records. Experiments with multiple microform catalogues.
1972–4	Internal production of subject index and periodical lists. Experiments to link Hatfield terminal to OCLC and UTLASS.
1975	Creation of BLAISE following online retrieval studies. Availability of retrospective MARC file on COM, provided by LASER.
1976–80	Use of LOCAS batch operation to produce complete catalogue for Hatfield and growth of a union catalogue for HERTIS Colleges.
1978	Development of Telepen stand-alone microprocessor based issue system.
1981	Conversion to SWALCAP online operation to provide integrated cataloguing and issue systems.

Fig 15 Chronology of main events

References

1. Command Paper 9703: *Technical Education*. Feb 1956, HMSO.
2. Williams, B J S *Evaluation of microreading techniques for information and data storage and retrieval*. (Final report to OSTI on a project funded from Aug 1965 to March 1967) HERTIS, 1967.
3. Wright, G H 'Reprography aids HERTIS in its bibliographical service to education and industry'. *Library Association Record* **67** July 1965, 215–221.
4. Bagley, D E *Computers in libraries: a select bibliography*. HERTIS, Hatfield College of Technology Library, 1966.
5. Higham, N *Computer needs for university library operations*. SCONUL, 1973.
6. Line, M B *Scope for ADP in the British Library: report for the Department of Education and Science*. HMSO, 1972.
7. Dufton, S P and Talbot R 'Computerised production of Dewey subject indexes for the libraries of Hertis'. *Program* **14** (1) 1980, 24–35.
8. 'Editing the BNB retrospective conversion file'. *British Library Bibliographical Services Division Newsletter* Aug 1976.
9. Holmes, P L *Online information retrieval: an introduction to the British Library Short Term Experimental Information Network project (STEIN)*. BLRDD Report 5360, 1977.
10. Duchesne, R M and Donbrowski L 'BNB/Brighton Public Libraries Catalogue project BRIMARC'. *Program* **7** Oct 1973, 205–224.
11. Ashford J, Bourne, R and Plaister J *Co-operation in library automation: the COLA project. Report of a LASER research project, 1974 – 1975*. London & South Eastern Library Region (LASER), 1975. (OSTI Report 5225).
12. Holmes, P L 'The British Library Automated Information Service, BLAISE', *Aslib Proceedings* **29** June 1977, 214–220.
13. Bryant, P, *Bath University Comparative Catalogue Study (BUCCS): final report*. 7 vols. Bath University Library, 1975.
14. 'BNB retrospective file'. *British Library Bibliographic Services Division Newsletter* May 1977.
15. Hopkinson, A 'MERLIN for the cataloguer'. *Aslib Proceedings* **29** Aug 1977, 284 – 294.
16. 'Hatfield Polytechnic Telepen system'. *VINE* **22** June 1978, 32–34.
17. Seal, A *Automated Cataloguing in the UK: a guide to services*. Bath University, Centre for Catalogue Research, 1980 (BLRDD Report 5545).
18. 'The Library co-operatives and LOCAS: ten years of growth'. *VINE* **28**, May 1979, 37 – 48.

19. 'Lancaster University Library, report of issue system problems'. *VINE* **14**, Dec 1975, 9.

20. Spencer, J R 'An appraisal of computer output microfilm for library catalogues'. National Reprographic Centre for documentation, 1974.

21. Reynolds, L *Visual presentation of information in COM library catalogues*. Royal College of Art, 1979. 2 vols. (BLRDD Report 5472).

22. Seal A, Bryant P, and Hall, C *Full and short entry catalogues*. Bath University, Centre for Catalogue Research, 1982. (BLRDD Report 5669).

23. 'Hatfield cataloguing output costs'. *VINE* **28** May 1979, 32–34.

24. King, M 'On costing alternative patterns for COM-fiche catalogues'. *Program* **14** Oct 1980, 147 – 160.

Glossary

AACR2	2nd edition of the Anglo American cataloguing rules.
AERE	Atomic Energy Research Establishment (UK Research Centre)
ALS	Automated Library Systems Limited – a company marketing equipment for automated circulation control systems.
BL	The British Library. Divisions include: BSD – Bibliographic Services Division. RDD – Research and Development Department.
BLAISE	British Library Automated Information SErvice – an automated service operated by BLBSD offering access to the MARC database and to various commercial databases for information retrieval and cataloguing purposes. Also offering a complete catalogue service (see LOCAS).
BLCMP	Birmingham Libraries Cooperative Mechanisation Project. Started as a BL research project to provide three libraries with individual catalogue services from a common computer database, using MARC records. Later developed to a widely available library cooperative service.
BNB	British National Bibliography. A database prepared from MARC records by the BL Bibliographical Services Divisions.
BUCCS	Bath University Comparative Catalogue Study. A research project sponsored by BLRDD to compare different types of library catalogue output. Later became Centre for Catalogue Research.

COLA	COoperation in Library Automation. A research study undertaken by LASER into the feasibility of library automation networking. 75
COM	Computer output microform – either microfilm or microfiche.
DES	Department of Education and Science. The UK government department ultimately providing finance for libraries in the public sector.
DOBIS	DOrtmund BIbliographic System. A specific computer database structure devised for the Library of Dortmund University, Germany.
EMMA	Extra-MARC Material: books and other material for which no record exists in the 'official' MARC databases. In UK terms it includes most pre 1950 imprints and non UK/US imprints amongst other things.
GEAC	A computer firm offering a complete software/ hardware package for library operations. Introduced in the UK in 1980
HERTIS	Hertfordshire Technical Library and Information Service. A generic term describing both the network of College libraries financed jointly and under the direction of the County Technical Librarian and the information service to industry supported from these resources.
ISBN	International Standard Book Number – a (theoretically) unique number assigned to each new book on publication. The 10-digit number indicates publisher and the language block in which the book was published, and contains a check-digit. The ISBN is used as key number in bibliographic databases.
LASER	London and South-Eastern Library Region – a regional co-operative established to facilitate inter-library lending and other co-operative activities.
LASSOS	Library Automation Systems and Services Overview Study. A committee set up by the BL Research and Development Department, changed in 1982 to Library Automation Group.
LC	Library of Congress. The US source of authoritative catalogue data in MARC format.

LOCAS   LOcal Cataloguing Service. A service provided
                by British Library Bibliographical Services Di-
                vision providing catalogue production on a
                batch basis.

MARC   MAchine Readable Cataloguing – a tagging
                scheme for bibliographic databases – used in
                the production of UK and US national biblio-
                graphies in machine-readable form (MARC
                tapes) and on the way to becoming the inter-
                national standard.

MERLIN   Machine Readable Library Information Net-
                work. A projected advanced computer system
                for British Library Bibliographic Services Divi-
                sion. Developed to a feasibility stage in 1977 but
                then shelved until a computer enhancement
                could be financed.

NRCd   National Reprographic Centre for documenta-
                tion. An international centre for advice and re-
                search into microforms and reprography
                funded by BL Research and Development De-
                partment, based at the Hatfield Polytechnic.

OCLC   Ohio College Library Center. (Now an interna-
                tional service entitled On-line Computer Library
                Center). A non-profit cooperative corporation
                based in Ohio with a very extensive data base
                of records. Offers both a catalogue service and
                software packages for other library operations.

OSTI   Office for Scientific and Technical Information,
                predecessor of BL Research and Development
                Department.

SBN   Standard Book Number. See ISBN

SCOLCAP   Scottish Libraries Cooperative Automation Pro-
                ject – a cooperative cataloguing project based in
                Edinburgh and offering a BLAISE/LOCAS ser-
                vice to libraries in North Britain.

SCONUL   Standing Conference of National and University
                Libraries.

STEIN   Short Term Experimental Information Network.
                The initial research project designed to test feas-
                ibility of on-line information retrieval in the UK.

SWALCAP   South Western Academic Libraries Cooperative
                Automation Project.
                Established as a BLRDD research project to test

	feasibility of on-line co-operative services to a local network. Developed like BLCMP into a non-profit agency.
Telepen	Automated circulation system manufactured by S.B. Electronic Systems Limited. Microprocessor controlled, using light pens and bar codes.
UGC	University Grants Commission. The funding body for UK Universities.
UKMARC	The British MARC tapes produced by BLBSD.
UTLAS	University of Toronto Library Automation Systems. An in-house computer which developed a catalogue support system capable of being extended to a network of libraries.
VDU	Visual Display Unit – a TV-type screen where data is displayed.

Note:

Acknowledgement is made to VINE for several of these definitions. VINE is a periodical produced four times a year by the BLRDD supported Information Office for Library Automation, based at Southampton University Library and is essential reading for anyone interested in the development of library automation in the UK.